WATCH ME PRACTICE

Math Workbook

Grade 3

Watch Me Practice Grade 3 Math Workbook

Copyright © 2021 by Belinda L. Spears

For more information about this title or to order other books and/or electronic media, contact the publisher:

B.L. Academic Services, LLC

http://watchmepracticeseries.com

info@watchmepracticeseries.com

First Edition

ISBN: 978-1-73692337-2-6

Printed in the United States

Cover Designer: Jamira Ink Designs LLC

Interior Designer/Editor: 1106 Design

Table of Contents

WRITING NUMBERS IN THE THOUSANDS

The number 7,654 has four digits. Each digit has a place value. For example:

The 7 is in the thousands place and is equal to 7,000

The 6 is in the hundreds place and is equal to 600

The 5 is in the tens place and is equal to 50

The 4 is in the ones place and is equal to 4

7,654 is written in standard form. The number 7,654 can also be written in words. **For example**: seven-thousand six hundred fifty-four.

Another way to help you read numbers in standard form through the thousands is by making a place-value chart. **For example:**

Thousands	Hundreds	Tens	Ones
7,	6	5	4

When writing a number in the thousands, always place a comma between the thousands place and the hundreds place.

Practice writing the four-digit numbers in a., b., and c. in standard form using a place-value chart below.

a. 9,876   b. 5,603   c. 8,065

Thousands	Hundreds	Tens	Ones

Practice writing the four-digit numbers in a., b., and c. in words on the lines below.

a. 9,876   b. 5,603   c. 8,065

Write the number in words:

a. _____

Write the number in words:

b. _____

Write the number in words:

c. _____

Create a four-digit number of your own. Write it in the place-value chart below.

Thousands	Hundreds	Tens	Ones

Now, write the four-digit number you created in words on the line below:

_____

WRITING NUMBERS IN THE TEN THOUSANDS

The number 54,321 has five digits. Each digit has a place value. **For example:**

The 5 is in the ten thousands place and is equal to 50,000

The 4 is in the thousands place and is equal to 4,000

The 3 is in the hundreds place and is equal to 300

The 2 is in the tens place and is equal to 20

The 1 is in the ones place and is equal to 1

54,321 is written in standard form. The number 54,321 can also be written in words.

For example: fifty-four thousand three hundred twenty-one.

Another way to help you read numbers in the ten thousands is by making a place-value chart.
For example:

Ten thousands	Thousands	Hundreds	Tens	Ones
5	4,	3	2	1

When writing numbers in the ten thousands, always place a comma between the thousands place and the hundreds place.

Practice writing the numbers in a., b., c., and d. in standard form using the place-value chart below.

a. 98,765   b. 56,402   c. 80,654   d. 32,456

Ten thousands	Thousands	Hundreds	Tens	Ones

Practice writing the numbers in a., b., c., and d. in words on the lines below.

Write the number in words:

a. _____

Write the number in words:

b. _____

Write the number in words:

c. _____

Write the number in words:

d. _____

Create a five-digit number of your own. Write it in the place-value chart below.

Ten thousands	Thousands	Hundreds	Tens	Ones

Now, write the five-digit number you created in words on the line below:

_____

WRITING NUMBERS IN THE HUNDRED THOUSANDS

The number 654,789 has six digits. Each digit has a place value. For example:

1. The 6 is in the hundred thousands place and is equal to 600,000
2. The 5 is in the ten thousands place and is equal to 50,000
3. The 4 is in the thousands place and is equal to 4,000
4. The 7 is in the hundreds place and is equal to 700
5. The 8 is in the tens place and is equal to 80
6. The 9 is in the ones place and is equal to 9

654,789 is written in standard form. The number 654,321 can also be written in words. For example: six-hundred fifty-four thousand three-hundred twenty-one.

Another way to help you read numbers in the hundred thousands is to make a place-value chart.
For example:

Hundred thousand	Ten thousand	Thousands	Hundreds	Tens	Ones
6	5	4,	3	2	1

Practice writing numbers in a., b., c., and d. in standard form using the place-value chart below.

a. 987,654   b. 564,321   c. 806,504   d. 324,567

Hundred thousands	Ten thousands	Thousands	Hundreds	Tens	Ones

Practice writing the numbers in a., b., c., and d. in words on the lines below.

Write the number in words:

a. _____

Write the number in words:

b. _____

Write the number in words:

c. _____

Write the number in words:

d. _____

Create a six-digit number of your own. Write it in the place-value chart below.

Hundred thousands	Ten thousands	Thousands	Hundreds	Tens	Ones

Now, write the 6-digit number you created in words on the line below:

_____

EXPANDED FORM

Writing Numbers in the Thousands in Expanded Form

The number 7,654 is written in standard form. Written in expanded form, the number 7,654 is written as:
7,000 + 600 + 50 + 4

Write the following numbers in expanded form.

    a. 9,876 _____

    b. 5,603 _____

    c. 8,065 _____

    d. 3,200 _____

Create a four-digit number and write it in three ways: standard form, words, and expanded form.

    1. Standard form: _____

    2. Words: _____

            _____

    3. Expanded form: _____

            _____

            _____

Writing Numbers in the Ten Thousands in Expanded Form

The number 54,321 is written in standard form. Written in expanded form, the number 54,321 is written as:
50,000 + 4,000 + 300 + 20 + 1

Write the following numbers in expanded form.

a. 98,765 _____

b. 56,402 _____

c. 80,654 _____

d. 32,456 _____

Create a five-digit number and write it in three ways: standard form, words, and expanded form.

a. Standard form: _____

b. Words: _____

_____

c. Expanded form: _____

_____

_____

Writing Numbers in the Hundred Thousands in Expanded Form

The number 654,321 is written in standard form. Written in expanded form, the number 654,321 is written as:
600,000 + 50,000 + 4,000 + 300 + 20 + 1

Write the following numbers in expanded form.

a. 987,654 _____

b. 564,321 _____

c. 806,504 _____

d. 324,567 _____

Create a six-digit number and write it in three ways: standard form, words, and expanded form.

1. Standard form: _____

2. Words: _____

_____

3. Expanded form: _____

_____

_____

GREATER THAN, LESS THAN, EQUAL TO

Comparing Numbers in the Thousands
Use the symbols >, <, or = to compare the following numbers:

1. 9,876 _____ 5,603

2. 3,200 _____ 9,876

3. 8,000 + 60 + 9 _____ 8,000 + 600 + 9

What Number Comes Before . . .

1. _____ , 9,876

2. _____ , 3,200

3. _____ , 8,065

What Number Comes After . . .

1. 8,069 _____

2. 5,603 _____

3. 9,257 _____

4. 2,032 _____

What Number Comes Between . . .

1. 9,876 and 9,878 _____

2. 3,200 and 3,202 _____

3. 4,555 and 4,557 _____

Comparing Numbers in the Ten Thousands
Use the symbols >, <, or = to compare the following numbers:

1. 80,654 _____ 98,765

2. 56,402 _____ 56,403

3. 20,000 + 3,000 + 700 + 80 + 9 _____ 20,000 + 3,000 + 700 + 80 + 9

What Number Comes Before . . .

1. _____ , 80,700

2. _____ , 56,400

3. _____ , 23,789

What Number Comes After . . .

1. 23,789 _____

2. 98,765 _____

3. 15,715 _____

4. 34,567 _____

What Number Comes Between . . .

1. 23,411 and 23,413 _____

2. 15,800 and 15,802 _____

3. 56,403 and 56,405 _____

Comparing Numbers in the Hundred Thousands

Use the symbols >, <, or = to compare the following numbers:

1. 564,321 _____ 324,567

2. 478,592 _____ 987,654

3. 800,000 + 6,000 + 500 + 4 _____ 800,000 + 6,000 + 500 + 5

What Number Comes Before . . .

1. _____ , 806,504

2. _____ , 324,567

3. _____ , 987,654

What Number Comes After . . .

1. 987,654 _____

2. 564,321 _____

3. 811,929 _____

What Number Comes Between . . .

1. 811,429 and 811,431 _____

2. 478,998 and 479,000 _____

3. 567,999 and 568,001 _____

ESTIMATING AND ROUNDING NUMBERS

Estimating and Rounding Numbers to the Nearest Ten

When rounding to the nearest ten, if the number in the ones place is between 1 and 4, round down to zero. If the number in the ones place is between 5 and 9, round the number in the tens place up to the next digit, and change the digit in the ones place to zero. For example:

To estimate the difference of 25 minus 14, round each number to the nearest ten.

25 becomes 30 because the digit in the ones place is between 5 and 9

14 becomes 10 because the digit in the ones place is between 1 and 4

The estimated difference of 25 minus 14 is 20 because 30−10 = 20.

Let's try another one. Estimate the sum of 51 and 37.

a. First, round 51 to the nearest ten _____

b. Next, round 37 to the nearest ten _____

The estimated sum of 51 plus 37 is 90 because _____ + _____ = 90

Estimate the sum of 23 plus 45

a. 23 rounded to the nearest ten becomes _____. **Explain why.**

_____

_____

b. 45 rounded to the nearest ten becomes _____. **Explain why.**

_____

_____

c. The estimated sum of 23 plus 45 is 70 _____. **Explain why.**

_____

_____

On your own, estimate the sum or difference of the following numbers:

1. 78 − 22 = _____

2. 15 + 30 = _____

Estimating and Rounding to the Nearest Hundred

When rounding to the nearest hundred, if the digit in the tens place is between 1 and 4, round down by changing the digit in the tens place and the digit in the ones place to zeros. **For example: 347 rounded to the nearest hundred becomes 300 because the digit in the tens place is between 1 and 4.**

If the digit in the tens place is between 5 and 9, round up by rounding the digit in the hundreds place up to the next digit. Then change the digits in the tens place and the ones place to zeros. **For example: 777 rounded to the nearest hundred becomes 800 because the digit in the tens place is between 5 and 9.**

Estimate the sum of 654 plus 321

a. First, round 654 to the nearest hundred _____

b. Next, round 321 to the nearest hundred _____

The estimated sum of 654 plus 321 is 1000 because _____ + _____ = 1000

Estimate the difference of 823 minus 479

a. 823 rounded to the nearest hundred becomes _____. **Explain why.**

_____

_____

b. 479 rounded to the nearest hundred becomes _____. **Explain why.**

_____

_____

c. The estimated difference of 823 minus 479 is _____. **Explain why.**

_____

_____

On your own, estimate the sum or difference between the following numbers:

1. $718 + 842 =$ _____

2. $562 - 450 =$ _____

Estimating and Rounding to the Nearest Thousand

When rounding to the nearest thousand, if the digit in the hundreds place is between 1 and 4, round down by changing the digits in the hundreds place, tens place, and ones place to zeros. **For example: 3,347 would become 3,000 because the digit in the hundreds place is between 1 and 4.**

If the digit in the hundreds place is between 5 and 9, round up by rounding the digit in the thousands place up to the next digit and changing the digits in the hundreds place, tens place, and ones place to zeros. **For example: 9,842 rounded to the nearest thousand becomes 10,000 because the digit in the hundreds place is between 5 and 9.**

Estimate the sum of 8,169 plus 2,718

    a.   First, round 8,169 to the nearest thousand _____

    b.   Next, round 2,718 to the nearest thousand _____

The estimated sum of 8,169 plus 2,718 is _____. **Explain why.**

_____

_____

_____

Estimate the difference of 5,987 minus 4,567

    a.   5,987 rounded to the nearest thousand becomes _____ .

    b.   4,567 rounded to the nearest thousand becomes _____ .

The estimated difference of 5,987 minus 4,567 is _____. **Explain why.**

_____

_____

_____

On your own, estimate the sum or difference of the following numbers:

    1. $3,210 + 7,654 =$ _____

    2. $8,930 - 6,543 =$ _____

WRITING ORDINAL NUMBERS THROUGH 100 BY SKIP COUNTING USING EVEN AND ODD NUMBERS

As you learned in Grade 2, ordinal numbers tell the position of numbers. Write the ordinal numbers and their abbreviations for Grade 2 through Grade 12.

Grade	Ordinal Number	Abbreviation
Grade 1	First Grade	1st Grade
Grade 2		
Grade 3		
Grade 4		
Grade 5		
Grade 6		
Grade 7		
Grade 8		
Grade 9		
Grade 10		
Grade 11		
Grade 12		

Starting at the number 14, skip count to 20 by twos using ordinal numbers and their abbreviations.

Number	Ordinal Number	Abbreviation
14		
20		

Starting at 20, skip count to 47 by threes using ordinal numbers and their abbreviations.

Number	Ordinal Number	Abbreviation
20		
47		

Starting at 48, skip count to 84 by fours using ordinal numbers and their abbreviations.

Number	Ordinal Number	Abbreviation
48		
84		

Starting at 85, skip count to 100 by fives using ordinal numbers and their abbreviations.

Number	Ordinal Number	Abbreviations
85		
100		

ADDITION

Adding Numbers with Three Digits

When adding numbers with three digits, add the numbers in the ones place first. Next, add the numbers in the tens place. Then, add the numbers in the hundreds place. Regroup, if necessary. For example:

$$
\begin{array}{r}
450 \\
+\ 562 \\
\hline
1,012
\end{array}
$$

Find the sums:

1.	842	2.	823	3.	555	4.	428
	+ 718		+ 479		+ 222		+ 354

Adding Numbers with Four Digits

When adding numbers with four digits, add the digits in the ones place first. Second, add the digits in the tens place. Third, add the digits in the hundreds place. Fourth, add the digits in the thousands place. Remember to place a comma between the thousands place and the hundreds place. Regroup, if necessary.

For example:

$$
\begin{array}{r}
8,169 \\
+\ 2,718 \\
\hline
10,887
\end{array}
$$

Find the sums:

1.	7,654	2.	5,728	3.	2,632	4.	4,576
	+ 3,210		+ 8,046		+ 1,547		+ 9,813

Adding Numbers with Five Digits

When adding numbers with five digits, add the digits in the ones place first. Second, add the digits in the tens place. Third, add the digits in the hundreds place. Fourth, add the digits in the thousands place. Fifth, add the digits in the ten thousands place. Regroup, if necessary.

For example:

```
   24,543
 + 30,714
 ───────
   55,257
```

Find the sums:

1.
```
    80,654
  + 98,725
  ───────
```

2.
```
    56,402
  + 56,403
  ───────
```

3.
```
    23,789
  + 15,710
  ───────
```

4.
```
    38,567
  + 44,342
  ───────
```

Adding Numbers with Six Digits

When adding numbers with six digits, add the digits in the ones place first. Second, add the digits in the tens place. Third, add the digits in the hundreds place. Fourth, add the digits in the thousands place. Fifth, add the digits in the ten thousands place. Sixth, add the digits in the hundred thousands place. Regroup if necessary.

For example:

```
   568,000
 + 429,112
 ────────
   997,112
```

Adding Numbers with Three Addends

When adding numbers with three addends, follow the same order as when adding numbers with two or more digits. Work from right to left, starting with the digits in the ones place. Regroup, if necessary.

For example

```
    312
      9
+ 1,437
-------
  1,758
```

Find the sums:

1.
```
      28
     140
+  3,507
-------
```

2.
```
    977
      2
+     0
------
```

3.
```
   6,543
     781
+      5
-------
```

Adding Numbers with Four Addends

When adding numbers with four addends, follow the same order as when adding numbers with two or more digits. Work from right to left, starting with the digits in the ones place. Regroup, if necessary.

For example:

```
    360
  1,285
     38
+     2
-------
  1,685
```

Find the sums.

1.
```
      28
     143
       2
+  3,567
-------
```

2.
```
    977
      2
     44
+     0
------
```

3.
```
   6,543
     781
      19
+      5
-------
```

SUBTRACTION

Subtracting Numbers with Three Digits

When subtracting numbers with three digits, you may have to regroup more than once. Subtract the digits in the ones place first. Next, subtract the digits in the tens place. Then, subtract the digits in the hundreds place. Regroup, if necessary, at each step.

For example:

```
    562
  − 473
  ─────
     89
```

Find the differences.

1.	842	2.	555	3.	454
	−718		−267		− 328

Subtracting Numbers with Four Digits

When subtracting numbers with four digits, you may have to regroup more than once. Subtract the digits in the ones place first. Next, subtract the digits in the tens place. Then, subtract the digits in the hundreds place. Last, subtract the digits in the thousands place. Regroup, if necessary, at each step.

For example:

```
   7,210
  −3,654
  ──────
   3,556
```

Follow the same order when subtracting numbers with five or six digits. Work from right to left. Start with the ones place first.

Find the differences.

1.	2,547 −1,639	**2.**	98,654 − 80,765	**3.**	811,429 − 478,998

MONEY

Adding and Subtracting Money

When adding or subtracting money amounts, follow the same order as when adding or subtracting numbers with two or more digits. Remember these important points:

1. Bring down the decimal.
2. The dollar amounts are on the left of the decimal point.
3. Cents are on the right of the decimal point.
4. Always place the dollar sign in front of the dollar amount.
5. Make sure the decimal points are lined up in each amount.
6. Regroup, if necessary.

For example:

$$\begin{array}{r} \$25.47 \\ + 16.39 \\ \hline \$41.86 \end{array} \qquad \begin{array}{r} \$32.55 \\ - 20.27 \\ \hline \$12.28 \end{array}$$

Find the sums or differences of the money amounts:

1.	$45.32 + 56.78	**2.**	$120.35 9.60 + 63.17	**3.**	$81.43 − 70.25	**4.**	$210.08 − 38.64

Money Word Problems

1. Tyler earned $10.50 mowing lawns. Tyler had $12.75 saved in his piggy bank. How much money does Tyler have in total?

2. Madison likes getting her nails painted. Madison received $15 this week for her allowance. It costs $7 to get her nails painted. If Madison spends $7 to get her nails painted, how much money will Madison have left?

MULTIPLICATION

In Grade 2, you learned that multiplication is a quick way of adding by skip counting. Skip counting can help you memorize multiplication facts. For example, when memorizing the multiplication facts of 2, remember to skip count by two.

$$2 \times 0 = 0$$
$$2 \times 1 = 2$$
$$2 \times 2 = 4$$
$$2 \times 3 = 6$$
$$2 \times 4 = 8$$
$$2 \times 5 = 10$$
$$2 \times 6 = 12$$
$$2 \times 7 = 14$$
$$2 \times 8 = 16$$
$$2 \times 9 = 18$$
$$2 \times 10 = 20$$
$$2 \times 11 = 22$$

Write the multiplication facts for the numbers 3 through 6:

3	4	5	6
3 x 0 = 0			

Write the multiplication facts for 7 through 10:

7	8	9	10
7 x 0 = 0			

Knowing your multiplication facts will help you when multiplying two-digit, three-digit, and four-digit numbers.

Multiplying Two-Digit Numbers by One-Digit Numbers

When multiplying two-digit numbers by one-digit numbers, multiply the one-digit number by the number in the ones column first. Then multiply the one-digit number by the number in the tens column. Regroup, if necessary.

For example:

$$
\begin{array}{r}
17 \\
\times 2 \\
\hline
34
\end{array}
$$
(two-digit number)
(one-digit number)

Multiply:

1.
$$
\begin{array}{r}
24 \\
\times 3 \\
\hline
\end{array}
$$

2.
$$
\begin{array}{r}
35 \\
\times 4 \\
\hline
\end{array}
$$

3.
$$
\begin{array}{r}
46 \\
\times 5 \\
\hline
\end{array}
$$

Multiplying Three-Digit Numbers by One-Digit Numbers

When multiplying three-digit numbers by one-digit numbers, multiply by the ones first. Second, multiply by the tens. Third, multiply by the hundreds.

For example:

$$
\begin{array}{r}
173 \\
\times 2 \\
\hline
346
\end{array}
$$
(three-digit number)
(one-digit number)

Multiply:

1.
$$
\begin{array}{r}
243 \\
\times 3 \\
\hline
\end{array}
$$

2.
$$
\begin{array}{r}
354 \\
\times 4 \\
\hline
\end{array}
$$

3.
$$
\begin{array}{r}
463 \\
\times 5 \\
\hline
\end{array}
$$

Multiplying Four-Digit Numbers by One-Digit Numbers

When multiplying four-digit numbers by one-digit numbers, multiply by the ones first. Second, multiply by the tens. Third, multiply by the hundreds. Fourth, multiply by the thousands.

For example:

$$
\begin{array}{r}
1,734 \\
\times\ 2 \\
\hline
3,468
\end{array}
$$
(four-digit number)
(one-digit number)

Multiply:

1.
$$
\begin{array}{r}
2,435 \\
\times\ 3 \\
\hline
\end{array}
$$

2.
$$
\begin{array}{r}
3,546 \\
\times\ 4 \\
\hline
\end{array}
$$

3.
$$
\begin{array}{r}
4,627 \\
\times\ 5 \\
\hline
\end{array}
$$

DIVISION

In Grade 2, you learned to think of division as a way of dividing objects into groups. Knowing your multiplication facts will help you do division. Division has several steps. Using a strategy will help you remember the steps. For example:

Drink

Milk

Strong

Bones

This strategy will help you remember the following steps when dividing:

D stands for **divide**

M stands for **multiply**

S stands for **subtract**

B stands for **bring down**

Dividing One-Digit Numbers by One-Digit Numbers

When dividing one-digit numbers by one-digit numbers, use the strategy **D**rink **M**ilk **S**trong **B**ones.

For example:

$$4\overline{)9}$$

1. **D**ivide the dividend 9 by the divisor 4. What number times 4 will equal 9 or come close to it? 4 x 2 = 8. Place the quotient 2 above the nine. Then go to the next step.

2. **M**ultiply the quotient 2 by the divisor 4 which gives a product of 8. Place the product 8 below the 9. Go to the next step.

3. **S**ubtract 8 from 9 to get a difference of 1.

*You would not go to the next step of "Bring Down," because there is no number to bring down when dividing one-digit numbers by one-digit numbers.

Therefore, $4\overline{)9}$ is 2 R1. This division problem can also be written as 9 ÷ 4 = 2 R1

Divide:

1. $5\overline{)5}$ **2.** $2\overline{)9}$ **3.** $3\overline{)8}$ **4.** $6\overline{)9}$

Dividing Two-Digit Numbers by One-Digit Numbers

When dividing two-digit numbers by one-digit numbers, work from left to right (using the numbers within the bracket) starting with the tens place. Then use the strategy **D**rink **M**ilk **S**trong **B**ones.

For example:

$$
\begin{array}{r}
14 \\
3\overline{\smash{)}42} \\
-3 \\
\hline
12 \\
-12 \\
\hline
0
\end{array}
$$

1. **D**ivide the tens. What number times 3 will equal 4 or come close to it? $3 \times 1 = 3$. Place the quotient 1 above the 4. Go to the next step.
2. **M**ultiply the quotient 1 by the divisor 3, which gives a product of 3. Place the product 3 below the 4. Go to the next step.
3. **S**ubtract 3 from 4 to get a difference of 1. Go to the next step.
4. **B**ring down the 2 and add it to the 1 for a total of 12. Start the steps from the beginning.
5. **D**ivide the new dividend of 12 by 3. What number times 3 will equal 12 or come close to it? $3 \times 4 = 12$. Place the quotient 4 above the 2. Then go to the next step.
6. **M**ultiply the quotient 4 by the divisor 3, which gives a product of 12. Place the product 12 below the 12. Go to the next step.
7. **S**ubtract 12 from 12 to get a difference of 0.

*Since there are no numbers left, you do not go on to the next step.

$3\overline{\smash{)}42}$ gives a quotient of 14 with no remainder. This division problem can also be written as $42 \div 3 = 14$.

*When dividing a two-digit number by a one-digit number, if the digit in the tens place is less than the divisor, divide the ones. For example: The digit (2) in the tens place is less than the divisor (6). So, divide the entire number, which is 24, by the divisor 6. The quotient (4) must be placed above the 4 in the ones column.

$$
\begin{array}{r}
4 \\
6\overline{\smash{)}24} \\
-24 \\
\hline
0
\end{array}
$$

Divide:

1. $5\overline{)25}$ **2.** $6\overline{)72}$ **3.** $4\overline{)34}$ **4.** $8\overline{)72}$

Dividing Three-Digit Numbers by One-Digit Numbers

When dividing three-digit numbers by one-digit numbers, work from left to right (using the numbers within the bracket) starting with the hundreds place. Then use the strategy **D**rink **M**ilk **S**trong **B**ones.

For example:

$$
\begin{array}{r}
81 \quad \text{R1} \\
3\overline{)244} \\
-24 \\
\hline
04 \\
-3 \\
\hline
1 \\
\end{array}
$$

1. **D**ivide the hundreds. The digit in the hundreds place is less than the divisor. So, divide the tens, which becomes 24. What number times 3 will equal 24 or come close to it? 8 x 3 = 24. Go to the next step.

2. **M**ultiply the quotient 8 by the divisor 3, which gives a product of 24. Place the product 24 below the 24. Go to the next step.

3. **S**ubtract 24 from 24 to get a difference of 0. Go to the next step.

4. **B**ring down the 4. Start the steps from the beginning.

5. **D**ivide the new dividend of 4 by 3. What number times 3 will equal 4 or come close to it? 3 x 1 = 3. Place the quotient 1 above the 4. Then go to the next step.

6. **M**ultiply the quotient 1 by the divisor 3, which gives a product of 3. Place the product 3 below the 4. Go to the next step.

7. **S**ubtract 3 from 4 to get a difference of 1.

*Since there are no numbers left, you do not go on to the next step.

$3\overline{)244}$ gives a quotient of 81, with a remainder of 1. This division problem can also be written as 244 ÷ 3 = 81 R1.

Divide:

1. $5\overline{)628}$
2. $7\overline{)377}$
3. $2\overline{)178}$
4. $8\overline{)492}$

FRACTIONS

In Grade 2, you learned that a fraction is made up of a numerator and a denominator. The numerator tells how much of the whole part is being used. The denominator gives the whole part. In the following example, tell which number is the numerator and which number is the denominator.

$$\frac{2}{4}$$

a. Which number is the numerator? _____

b. Which number is the denominator? _____

Fractions that have the same denominator are called like fractions. Fractions that have different denominators are called unlike fractions. When both the numerator and the denominator are the same number, it is equal to the whole number 1.

Adding Fractions with the Same Denominator

When adding fractions with the same denominator, add the numerators. The denominators stay the same. **For example:**

$$\frac{2}{4} + \frac{1}{4} = \frac{3}{4}$$

Add.

1. $\frac{3}{7} + \frac{3}{7}$

2. $\frac{4}{8} + \frac{2}{8}$

3. $\frac{5}{9} + \frac{4}{9}$

4. $\frac{3}{10} + \frac{6}{10}$

Subtracting Fractions with the Same Denominator

When subtracting fractions with the same denominator, subtract the numerators.
The denominators stay the same.

For example:

$$\begin{array}{r} \dfrac{2}{4} \\ -\ \dfrac{1}{4} \\ \hline =\ \dfrac{1}{4} \end{array}$$

Subtract.

1.
$$\begin{array}{r} \dfrac{3}{7} \\ -\ \dfrac{2}{7} \\ \hline \end{array}$$

2.
$$\begin{array}{r} \dfrac{4}{8} \\ -\ \dfrac{2}{8} \\ \hline \end{array}$$

3.
$$\begin{array}{r} \dfrac{5}{9} \\ -\ \dfrac{3}{9} \\ \hline \end{array}$$

4.
$$\begin{array}{r} \dfrac{6}{10} \\ -\ \dfrac{3}{10} \\ \hline \end{array}$$

ROMAN NUMERALS

In Grade 2, you learned that Roman numerals are numbers written in the Latin language. The ancient Romans spoke Latin. Here are the Roman numerals for numbers 1 through 20.

ROMAN NUMERALS	NUMBERS
I	1
II	2
III	3
IV	4
V	5
VI	6
VII	7
VIII	8
IX	9
X	10
XI	11
XII	12
XIII	13
XIV	14
XV	15
XVI	16
XVII	17
XVIII	18
XIX	19
XX	20

Add or subtract the Roman numerals.

1. IV + V = _____

2. X + II = _____

3. XIV + VI = _____

4. XX – X = _____

5. XIII – V = _____

6. XVIII – IX = _____

WORD PROBLEMS

Addition

1. Robert lives in a two-story home. There are 12 stairs leading to the basement. There are 15 stairs leading to the 2nd floor. How many stairs are in Robert's home? _____

2. At the pet store, two cats gave birth to kittens. There were 6 kittens born to the first cat. Seven kittens were born to the second cat. On the same day, two kittens were donated by Miss Imani, and Jake from across the lake donated 10 kittens! How many kittens did the pet store receive on that day?

Subtraction

1. On the Blueberry Farm, there are 157 trees. Mr. Bart, the owner of Blueberry Farm, donated 48 trees to the Apple Cider Farm. How many trees did Mr. Bart have left? _____

2. The Silly Saver Supermarket placed a case of 120 cans of tomato sauce outside of their store at 8 o'clock in the morning. The weatherman predicted strong afternoon storms. The storms knocked out power to many homes. The strong winds knocked over 68 cans of the tomato sauce. How many cans were left standing? _____

Multiplication

1. Emily and Tashauna like to hang decorations on their Christmas tree. Tashauna hung 8 ornaments on the tree. Emily hung 3 times as many ornaments. How many more ornaments did Emily hang on her Christmas tree?

2. Eldon and Jayson rode their bikes on a sunny afternoon. Eldon rode 12 miles. Jayson rode four times as many miles. How many more miles did Jayson ride? _____